ENERGY SECTOR STANDARD
OF THE PEOPLE'S REPUBLIC OF CHINA

中华人民共和国能源行业标准

Code for Planning and Design of Landscape
for Hydropower Projects

水电工程景观规划设计规范

NB/T 10346-2019

Chief Development Department: China Renewable Energy Engineering Institute
Approval Department: National Energy Administration of the People's Republic of China
Implementation Date: July 1, 2020

China Water & Power Press
中国水利水电出版社
Beijing 2024

All rights reserved. No part of this publication may be reproduced, stored in a retrieval system, or transmitted in any form or by any means—electronic, mechanical, photocopying, recording or otherwise, without prior written permission of the publisher.

图书在版编目（CIP）数据

水电工程景观规划设计规范 ：NB/T 10346—2019 = Code for Planning and Design of Landscape for Hydropower Projects (NB/T 10346—2019) ：英文 / 国家能源局发布. -- 北京 ：中国水利水电出版社, 2024. 6. -- ISBN 978-7-5226-2596-6

Ⅰ. TV-65

中国国家版本馆CIP数据核字第202484ZY74号

ENERGY SECTOR STANDARD
OF THE PEOPLE'S REPUBLIC OF CHINA
中华人民共和国能源行业标准

Code for Planning and Design of Landscape
for Hydropower Projects
水电工程景观规划设计规范
NB/T 10346-2019
（英文版）

Issued by National Energy Administration of the People's Republic of China
国家能源局　发布
Translation organized by China Renewable Energy Engineering Institute
水电水利规划设计总院　组织翻译
Published by China Water & Power Press
中国水利水电出版社　出版发行
　　Tel: (+ 86 10) 68545888　68545874
　　sales@mwr.gov.cn
　　Account name: China Water & Power Press
　　Address: No.1, Yuyuantan Nanlu, Haidian District, Beijing 100038, China
　　http：//www.waterpub.com.cn
中国水利水电出版社微机排版中心　排版
北京中献拓方科技发展有限公司　印刷
184mm×260mm　16开本　3印张　95千字
2024 年 6 月第 1 版　2024 年 6 月第 1 次印刷

Price（定价）：￥490.00

Introduction

This English version is one of China's energy sector standard series in English. Its translation was organized by China Renewable Energy Engineering Institute authorized by National Energy Administration of the People's Republic of China in compliance with relevant procedures and stipulations. This English version was issued by National Energy Administration of the People's Republic of China in Announcement [2023] No. 8 dated December 28, 2023.

This version was translated from the Chinese Standard NB/T 10346-2019, *Code for Planning and Design of Landscape for Hydropower Projects*, published by China Water & Power Press. The copyright is reserved by National Energy Administration of the People's Republic of China. In the event of any discrepancy in the implementation, the Chinese version shall prevail.

Many thanks go to the staff from the relevant standard development organizations and those who have provided generous assistance in the translation and review process.

For further improvement of the English version, any comments and suggestions are welcome and should be addressed to:

China Renewable Energy Engineering Institute
No. 2 Beixiaojie, Liupukang, Xicheng District, Beijing 100120, China
Website: www.creei.cn

Translating organizations:

China Renewable Energy Engineering Institute

POWERCHINA Beijing Engineering Corporation Limited

Translating staff:

WANG Meng	JIN Yi	GAO Yan	LU Bo
ZHANG Liying	ZHAO Fan	ZHU Yunuo	XIA Tong
QI Wen	MA Shijun	LIU Ruoqi	

Review panel members:

QIE Chunsheng	Senior English Translator
LI Zhongjie	POWERCHINA Northwest Engineering Corporation Limited
LI Kejia	POWERCHINA Northwest Engineering Corporation Limited

CHEN Lei	POWERCHINA Zhongnan Engineering Corporation Limited
ZHANG Qian	POWERCHINA Guiyang Engineering Corporation Limited
YAN Wenjun	Army Academy of Armored Forces, PLA
ZHANG Ming	Tsinghua University
YU Weiqi	China Renewable Energy Engineering Institute

National Energy Administration of the People's Republic of China

翻译出版说明

本译本为国家能源局委托水电水利规划设计总院按照有关程序和规定，统一组织翻译的能源行业标准英文版系列译本之一。2023 年 12 月 28 日，国家能源局以 2023 年第 8 号公告予以公布。

本译本是根据中国水利水电出版社出版的《水电工程景观规划设计规范》NB/T 10346—2019 翻译的，著作权归国家能源局所有。在使用过程中，如出现异议，以中文版为准。

本译本在翻译和审核过程中，本标准编制单位及编制组有关成员给予了积极协助。

为不断提高本译本的质量，欢迎使用者提出意见和建议，并反馈给水电水利规划设计总院。

地址：北京市西城区六铺炕北小街 2 号
邮编：100120
网址：www.creei.cn

本译本翻译单位：水电水利规划设计总院
　　　　　　　　中国电建集团北京勘测设计研究院有限公司
本译本翻译人员：王　猛　金　弈　高　燕　陆　波
　　　　　　　　张莉颖　赵　凡　朱雨诺　夏　彤
　　　　　　　　齐　文　马世军　刘若琦

本译本审核人员：
　　郄春生　英语高级翻译
　　李仲杰　中国电建集团西北勘测设计研究院有限公司
　　李可佳　中国电建集团西北勘测设计研究院有限公司
　　陈　蕾　中国电建集团中南勘测设计研究院有限公司
　　张　倩　中国电建集团贵阳勘测设计研究院有限公司
　　闫文军　中国人民解放军陆军装甲兵学院
　　张　明　清华大学
　　喻卫奇　水电水利规划设计总院

国家能源局

Announcement of National Energy Administration of the People's Republic of China [2019] No. 8

National Energy Administration of the People's Republic of China has approved and issued 152 energy sector standards including *Code for Operating and Overhauling of Excitation System of Small Hydropower Units* (Attachment 1), and the English version of 39 energy sector standards including *Code for Safe and Civilized Construction of Onshore Wind Power Projects* (Attachment 2).

Attachments: 1. Directory of Sector Standards
2. Directory of English Version of Sector Standards

National Energy Administration of the People's Republic of China

December 30, 2019

Attachment 1:

Directory of Sector Standards

Serial number	Standard No.	Title	Replaced standard No.	Adopted international standard No.	Approval date	Implementation date
...						
23	NB/T 10346-2019	Code for Planning and Design of Landscape for Hydropower Projects			2019-12-30	2020-07-01
...						

Foreword

According to the requirements of Document GNKJ [2015] No. 283 issued by National Energy Administration of the People's Republic of China, "Notice on Releasing the Development and Revision Plan of the Energy Sector Standards in 2015", and after extensive investigation and research, summarization of practical experience, and wide solicitation of opinions, the drafting group has prepared this code.

The main technical contents of this code include: basic requirements, basic data, landscape planning, landscape design, and costs.

National Energy Administration of the People's Republic of China is in charge of the administration of this code. China Renewable Energy Engineering Institute has proposed this code and is responsible for its routine management. Energy Sector Standardization Technical Committee on Hydropower Planning, Resettlement and Environmental Protection is responsible for the explanation of specific technical contents. Comments and suggestions in the implementation of this code should be addressed to:

China Renewable Energy Engineering Institute
No. 2 Beixiaojie, Liupukang, Xicheng District, Beijing 100120, China

Chief development organizations:

China Renewable Energy Engineering Institute

POWERCHINA Beijing Engineering Corporation Limited

Participating development organizations:

POWERCHINA Huadong Engineering Corporation Limited

POWERCHINA Chengdu Engineering Corporation Limited

POWERCHINA Zhongnan Engineering Corporation Limited

POWERCHINA Kunming Engineering Corporation Limited

POWERCHINA Guiyang Engineering Corporation Limited

Chief drafting staff:

ZHANG Liying	YU Weiqi	ZHAO Fan	JIN Yi
LIU Yuhan	FU Rui	WU Wenyou	YANG Wenzheng
XIA Quan	ZHOU Yun	WANG Binbin	LU Bo
ZOU Binghua	ZHANG Xian	JIANG Lihong	SHAN Chengkang

LI Qianqian	GU Jiajie	XIA Tong	TAN Qilin
MA Shijun	WANG Yushuang	HE Yuxuan	ZHANG Xu
TANG Huang	SUN Yuan		

Review panel members:

WAN Wengong	CHEN Guozhu	LI Yanong	CHEN Yongbai
ZENG Deyong	SHU Zeping	HE Xueming	CHEN Yuying
WANG Xingtai	QIU Jinsheng	LIU Guihua	GAO Liang
ZHOU Bo	CHEN Wenhua	LI Shisheng	

Contents

1	**General Provisions**	**1**
2	**Terms**	**2**
3	**Basic Requirements**	**3**
4	**Basic Data**	**5**
4.1	General Requirements	5
4.2	Project Data	5
4.3	Natural Environment Data	5
4.4	Social Environment Data	6
4.5	Landscape Data	6
4.6	Data Analysis	6
5	**Landscape Planning**	**7**
5.1	General Requirements	7
5.2	Guiding Philosophy and Principles	7
5.3	Planning Objectives	7
5.4	Planning Indicators	8
5.5	Planning Layout	8
5.6	Zone Planning	8
5.7	Special Planning	11
5.8	Planning Outcomes	12
6	**Landscape Design**	**14**
6.1	General Requirements	14
6.2	Geomorphologic Landscape Design	14
6.3	Landscape Design for Main Works of Hydropower Project	15
6.4	Design of Landscape Architectures	18
6.5	Design of Environmental Facilities	18
6.6	Planting Design	19
6.7	Landscape Lighting Design	21
6.8	Design of Auxiliary Facilities	21
6.9	Design Outcomes	22
7	**Costs**	**24**
Appendix A	**Indicators and Evaluation Standards for Landscape Planning of Hydropower Projects**	**25**
Appendix B	**Contents of Landscape Planning Report for Hydropower Projects**	**34**
Explanation of Wording in This Code		**35**
List of Quoted Standards		**36**

1 General Provisions

1.0.1 This code is formulated with a view to standardizing the content, methods and technical requirements of planning and design of landscape for hydropower projects.

1.0.2 This code is applicable to the planning and design of landscape for hydropower projects.

1.0.3 The planning and design of landscape for a hydropower project shall be determined according to the development purpose of the hydropower project or the protection and development requirements of regional landscape resources to embody the needs for the protection and development of landscape resources of the hydropower project, and should be coordinated with local tourism planning and follow the principles of aesthetics, applicability, safety, cost-effectiveness and sustainability. When landscaping is required in the hydropower project mission statement, special demonstration and design shall be carried out.

1.0.4 In addition to this code, the planning and design of landscape for hydropower projects shall comply with other current relevant standards of China.

2 Terms

2.0.1 landscape planning

planning of visual landscape image, and environment, ecology, and greening within a specific geographical area, based on the regional characteristics and attributes and taking into account the public psychology and behavior

2.0.2 landscape design

process of creating a beautiful environment and a recreational space within a specific geographical area through gardening techniques and engineering, by means of terrain shaping, planting, buildings and garden structures construction, and garden path arrangement

2.0.3 landscape pattern

spatial distribution and combination of landscape units of different sizes, shapes and attributes

2.0.4 landscape nodes

scenery spots with multiple functions within a specific geographical area

2.0.5 landscape elements

major landscape components such as terrain, water body, plants, buildings, squares and paths, garden ornaments, etc.

3 Basic Requirements

3.0.1 The landscape planning and design of a hydropower project shall, based on data collection, carry out the overall planning, zone planning, special planning, and landscape design for the project area through comparison of alternatives according to the protection and development objectives of the project landscape resources.

3.0.2 The landscape planning and design shall cover the permanent requisitioned land, temporarily occupied land and resettlement areas for the project. Peripheral areas that have a significant impact on the landscape and environmental features of the main works of the power station should be included in the landscape study.

3.0.3 The landscape planning and design shall, based on the natural landscape texture and the main works of the power station, employ the natural landscape as the main body, embody the industrial and technological features, take into account the leisure, recreation and themed tourism, and show the regional culture, so as to integrate the hydropower project into the natural landscape.

3.0.4 The landscape planning and design shall take full advantage of landscape resources, respect and protect natural and cultural heritages, understand and promote local cultural features, rationally utilize landscape components, so as to shape a characteristic landscape through arrangement of landscape elements.

3.0.5 The landscape planning and design shall, based on comparison of alternatives, carry out a typical design for key areas such as the hydropower complex and operation management camp.

3.0.6 The landscape planning should be carried out at the feasibility study stage of the hydropower project. The landscape planning shall define the landscape features and determine the overall landscape layout, functional zones, and landscape nodes according to the regional landscape pattern and the characteristics of the hydropower project. The landscape functional zoning shall be conducted according to the landscape type, purpose of space or activity. The zone planning shall take into account the layout of main works to determine the landscape layout of the hydropower project.

3.0.7 The landscape design should be carried out at the construction stage of main works, and be implemented by steps according the construction progress of the hydropower project. The landscape design shall be divided, according to the level of detail, into the scheme design, preliminary design and construction drawing design, and shall meet the following requirements:

1 The scheme design of landscape shall, based on the landscape planning, carry out the design of landscape function, style and features, traffic streamline, spatial relationship, and plant arrangement, and prepare the estimate.

2 The preliminary design of landscape shall, based on the scheme design, carry out the design of general layout, vertical layout, planting, waterscape, landscape architecture, environmental facilities, and pavement, and prepare the budget estimate.

3 The construction drawing design of the landscape shall, based on the preliminary design, determine the construction locations, dimensions and methods for the landscape works, and shall meet the requirements of construction and budget preparation.

3.0.8 The outcomes of landscape planning and design for a hydropower project shall include a report and drawings. The report shall be complete, well organized and concise, with a clear conclusion; the drawings shall be consistent with the text, with clear content, accurate expression, and standardized labeling.

4 Basic Data

4.1 General Requirements

4.1.1 The landscape planning and design of a hydropower project shall collect the project data and information on the natural environment, social environment and landscape in the project area.

4.1.2 The landscape planning and design of hydropower projects shall consider landscape resource evaluation and environmental investigation and evaluation.

4.2 Project Data

4.2.1 The landscape planning and design of a hydropower project shall collect the project data such as the planning condition, project condition and location condition of the hydropower project.

4.2.2 The planning conditions shall include the boundary line of land, total area, topographic map, locations of the external entrances and surrounding environment. The scale of the topographic map shall meet the following requirements:

    1   For the hydropower complex area, 1 : 500 to 1 : 2 000.

    2   For the reservoir area, 1 : 2 000 to 1 : 10 000.

    3   For roads, construction sites, etc., 1 : 500 to 1 : 2 000.

    4   For the specific landscape nodes, 1 : 500 to 1 : 2 000.

4.2.3 The project conditions shall include the project layout, construction layout, construction land requisition and implementation plan.

4.2.4 The location conditions shall include the geographic location, transportation facilities and surrounding condition.

4.3 Natural Environment Data

4.3.1 The landscape planning and design of a hydropower project shall collect the data on meteorology, geomorphology, geology, soil, biology, etc. in the project area.

4.3.2 The meteorological data shall include the temperature, humidity, precipitation, evaporation, wind direction, wind speed and sunshine of the representative meteorological station.

4.3.3 The geomorphological and geological data shall include the landform, altitude, slope, geological structure, geological hazards and groundwater level.

4.3.4 The soil data shall include the soil types and properties.

4.3.5 The biological data shall include the vegetation, biological resources and rare species.

4.4 Social Environment Data

4.4.1 The landscape planning and design of a hydropower project shall collect social environment data such as social history, culture, economy and planning of the project area.

4.4.2 The social history data shall include the historical evolution and protected historical objects.

4.4.3 The cultural data shall include traditional culture, ethnic culture, regional culture and corporate culture.

4.4.4 The economic data shall include the local GDP, residents' income, fiscal revenue, industrial structure, local specific economy and products, and public utilities.

4.4.5 The planning data shall include land use planning, urban and rural planning and traffic planning.

4.5 Landscape Data

The landscape planning and design of a hydropower project shall collect the scenery data on historical sites and relics, buildings and facilities, tourism resource evaluation, tourism products and route planning, and key item planning.

4.6 Data Analysis

The landscape planning and design of a hydropower project shall compile and analyze the collected data, state the sources of the data, analyze the reliability of the data, and make reasonable use of the data.

5 Landscape Planning

5.1 General Requirements

5.1.1 The landscape planning of a hydropower project shall, according to the conditions of project area, determine the requirements of protection and development of landscape resources, set reasonable objectives, coordinate the natural landscape and industrial landscape, and determine the landscape style of the hydropower project.

5.1.2 The landscape planning of a hydropower project shall include guiding philosophy and principles, planning objectives, planning indicators, planning layout, zone planning, special planning, etc.

5.2 Guiding Philosophy and Principles

5.2.1 The landscape planning shall, based on the general layout of the hydropower project, analyze and evaluate the current conditions of regional environment and project site, and consider the needs of interested parties, to formulate the planning scheme.

5.2.2 The landscape planning shall follow the principles below:

1. Adaptability to local conditions. The landscape planning shall make full use of the existing landform and embody local characteristics.

2. Ecology-based. The landscape planning shall be based on ecological protection and restoration, and create a harmonious relationship between the hydropower project and the natural environment.

3. Multifunctional. The landscape planning shall be carried out in combination with the functional zones of the hydropower project, and meet the diverse usage requirements in terms of safety, production, and living.

4. People-oriented. The landscape planning shall focus on improving the living environment and include the humanistic needs.

5. Accentuation. The landscape planning shall highlight the industrial attributes of the main buildings, and present the corporate image in the operation management camp and other important landscape nodes.

5.3 Planning Objectives

The landscape planning of a hydropower project shall determine planning objectives according to the landscape protection requirements and the development purpose of landscape resources, to improve the quality of living environment and create a "one project, one style" landscape considering the tourism development.

5.4 Planning Indicators

5.4.1 The landscape planning of a hydropower project shall, according to the landscape planning objectives, determine landscape planning indicators in terms of landscape resource value, development and utilization condition, and planning effect.

5.4.2 The landscape planning of a hydropower project should formulate a landscape planning indicator system in different levels and conduct value assignment and evaluation based on the site investigation, data collection, and relevant planning, to determine the landscape grade of the hydropower project. The indicators and evaluation standards for landscape planning of hydropower projects shall be in accordance with Appendix A of this code.

5.5 Planning Layout

5.5.1 The landscape planning layout of a hydropower project shall meet the following requirements:

1. The planning layout shall follow the principles of integrity and uniformity, clear priorities, adaptability to local conditions, combining landscape nodes with lines, combining lines with areas, and the reasonable combination of nodes, lines and areas.

2. The landscape zoning shall define the spatial boundaries according to the landscape type, purpose of space or activity.

3. Each zone shall have its characteristics in terms of activity type and facilities.

5.5.2 The landscape planning of a hydropower project shall cover the landscape areas of hydropower complex, reservoir, operation management camp, image display, road greening, temporary land occupation and resettlement according to the project layout, and shall define the key planning area and determine the overall layout plan according to the characteristics of each zone.

5.6 Zone Planning

5.6.1 The landscape zone planning for a hydropower project shall meet the following requirements:

1. Define the functions, landscape features and construction indicators of each zone.

2   Determine the layout plan of primary and secondary nodes according to the scale of each zone.

5.6.2   For the hydropower complex area, the planning shall meet the following requirements:

1   The landscape area of hydropower complex shall include dam, surface powerhouse, etc.

2   The landscape planning of dam shall include recommendation on dam type, beautification of dam crest, downstream slope decoration of dam body, slope greening of dam abutment, intake and other complex structures, as well as planning of viewing point near the dam abutments and recommendation on ancillary architectural style.

3   The landscape planning of surface powerhouse shall include the recommendation of powerhouse architectural style and the landscape planning of the powerhouse area.

5.6.3   For the reservoir area, the planning shall meet the following requirements:

1   The reservoir landscape area shall include waters, islands, area upstream of dam, drawdown area, and reservoir bank slopes and gentle hills within the range of the permanent requisitioned land.

2   The area upstream of dam and the reservoir bank slopes and gentle hills within the range of the permanent requisitioned land should be subjected to unified landscape planning, and the layout planning of landscape nodes should be conducted according to the overall scale. The key construction items should include revetments, waterfront platforms, recreational and sightseeing spaces, and other cultural and resort attractions based on the hydropower project.

3   The landscape planning of the drawdown area may be carried out selectively according to the difference in surrounding environment, use various methods for beautification, and make the area harmonious with the reservoir area.

5.6.4   For the operation management camp, the planning shall meet the following requirements:

1   Aim to create a natural, harmonious, beautiful and comfortable living and working environment.

2   Preliminarily determine the general layout and architectural style of buildings and the landscape features, and the architectural style should

embody the corporate culture.

3 Plan a certain space for evacuation square and fitness and sports facilities, and consider the setting of waterscape according to the site conditions.

4 Separate quiet zones from less ones according to the functional characteristic of the buildings.

5.6.5 For the image display area, the planning shall meet the following requirements:

1 The image display area should include exhibition area, image area at entrance, fish restocking station, etc.

2 The landscape planning of the image display area shall display the image and feature of the hydropower project, focusing on hydropower technology and corporate culture.

5.6.6 For the road greening area, the planning shall meet the following requirements:

1 The road greening area shall include the road to dam, access road, road around reservoir, etc.

2 Landscape planning shall be carried out for the diverse roadside space of different roads, and landscape nodes shall be set for tunnel portal, access tunnel portal of underground powerhouse, permanent bridge, and roadside areas with good vision, according to the actual needs of site.

5.6.7 For the temporarily occupied land, the planning shall meet the following requirements:

1 The temporarily occupied land should include spoil area, quarry and borrow area, land for construction, etc.

2 The temporarily occupied lands shall be classified and planned according to the disturbance and the site conditions. For temporarily occupied areas of high value, landscape nodes shall be appropriately planned considering the subsequent reclamation.

3 The landscape requirements for temporarily occupied land in different periods shall be presented with emphasis on time sequence, to ensure that the landscape of temporarily occupied area meets the environmental requirements of the hydropower project in construction period.

5.6.8 For the resettlement area, the planning shall meet the following requirements:

1. Build a resettlement area that is livable and touristy, with delightful environment and outstanding regional and cultural features.

2. Determine the landscape layout and landscape features of resettlement site, and recommend architectural style.

5.7 Special Planning

5.7.1 The landscape special planning of hydropower projects shall be carried out on the basis of the landscape zone planning. Landscape special planning shall include plant arrangement planning, landscape lighting planning, traffic streamline planning, tourism development planning and construction sequence planning.

5.7.2 The plant arrangement planning shall meet the following requirements:

1. The planning shall put forward the objectives and ideas of plant arrangement according to the characteristics of plants in the hydropower project area and the characteristics of landscape functional zones.

2. The planning shall balance the actual needs of ecology, production and living according to the ecological restoration of hydropower projects, and should divide the plant landscape into three types, i.e. ecological restoration, production protection and living enrichment, and propose corresponding plant arrangement measures.

3. For the plant landscape for living enrichment, the planning shall propose the characteristics of plant arrangement in atmosphere creation according to the functional requirement of landscape nodes.

4. The planning shall make the plant arrangement scheme for project area using the trees and grasses of local species.

5.7.3 Landscape lighting planning shall meet the following requirements:

1. The planning shall meet the safety needs of each area and make a rational night lighting design.

2. The planning should, based on the landscape functional zoning, consider ornamental and decorative lighting in addition to the basic lighting.

3. For the operation management camps, dam and other landmarks in the hydropower complex area, the planning should add special and festival lighting as appropriate.

5.7.4 The traffic streamline planning shall meet the following requirements:

1. The hydropower project area shall be classified for traffic streamlining, and a multi-level garden path system shall be planned using the roadway system.

2. Recreation garden path should be arranged in a ring.

3. Operation management camps and such areas should have the pedestrian system separated from vehicle system.

4. A rational parking capacity shall be planned considering the needs of staff, visitors and future tourism development.

5. The planning shall select centralized or decentralized parking facilities considering the actual needs of landscape functional zones and main works.

5.7.5 The tourism development planning shall meet the following requirements:

1. The planning shall, based on the attractions distribution, tourist number and other tourism-related conditions in the hydropower project area, analyze and determine the type of tourist market and predict the scale of tourism development.

2. The planning shall, based on the objectives in the hydropower project landscape planning, present the tourism development goal and orientation for the hydropower project considering the actual situation.

3. The planning shall include the characteristic attractions and tour routes for the hydropower project.

4. The planning shall include service facilities for the hydropower project.

5.7.6 The time sequence planning for landscape construction shall propose a reasonable schedule according to the construction sequence of the main works.

5.8 Planning Outcomes

5.8.1 A landscape planning report for the hydropower project shall be prepared, which should include planning background, current situation analysis, planning basis and principle, planning objective and concept, planning scheme, special planning, and cost estimate. The contents of landscape planning report for hydropower projects should be in accordance with Appendix B of this code.

5.8.2 The landscape planning for the hydropower project shall provide the following tables:

1   Indicators of landscape planning for the hydropower project.

2   Effect evaluation of landscape planning for the hydropower project.

5.8.3 The landscape planning for the hydropower project shall propose the following drawings:

1   Location map of landscape planning.

2   General layout of landscape planning.

3   Landscape functional zoning map.

4   Layout of zone planning.

5   Layout and renderings of important landscape nodes.

6 Landscape Design

6.1 General Requirements

6.1.1 The landscape design of a hydropower project shall, based on the general requirements of landscape planning and geomorphologic landscape design, carry out the main works landscape design, landscape architecture design, environmental facility design, planting design, landscape lighting design and auxiliary facility design of the hydropower project. The landscape designs shall be in harmony with each other.

6.1.2 The landscape design of a hydropower project shall adopt the method of combining key design with typical design.

6.2 Geomorphologic Landscape Design

6.2.1 The geomorphologic landscape design shall fully understand and summarize the local natural landform characteristics, and use remote sensing interpretation and other means to form the basic data on geomorphologic landscape resources serving as important basic conditions for project landscape design. The basic conditions of project landscape design should include the following:

1. Field sample area investigation and data collection.
2. Field survey and quantitative description.
3. Typical landform modeling and 3D simulation.
4. Remote sensing dynamic monitoring.
5. Suggestions on the development of the planned area and evaluation conclusion.

6.2.2 Geomorphologic landscape design shall make classification and assessment on the major landform changes taking into account the project planning and layout, and take appropriate measures. The geomorphologic landscape design shall meet the following requirements:

1. Ecological restoration shall be carried out along the road and in the slope area of the dam abutment.
2. Soil and water conservation restoration shall be carried out for the spoil area.
3. The living camp area shall be appropriately landscaped to be integrated into the natural environment.
4. Production buildings and structures such as dam and powerhouse shall

be appropriately beautified.

6.2.3 The typical design of geomorphologic landscape shall meet the following requirements:

1 Determine the suitable construction area, restricted construction area and prohibited construction area for the geomorphologic landscape from the perspective of the full construction lifecycle of hydropower projects, based on the 3D visualized modeling of terrain.

2 Propose the principles and indicators of landscape construction from the perspective of eco-environment protection.

3 Propose the content and form of landscape construction taking into account the local and regional culture and deliver landscape design outcomes.

6.2.4 The outcomes of landform landscape design shall include the following:

1 3D models of landscape design, including 3D models of typical design area landscapes.

2 Dynamic display of landscape effects.

6.3 Landscape Design for Main Works of Hydropower Project

6.3.1 The landscape design of the dam complex shall meet the following requirements:

1 The selection of dam type should consider landscaping, and recommendation should be given to architectural styles of associated buildings taking into account the landscape features of the hydropower project.

2 The dam crest beautification design shall include dam crest railings, lamp posts, dam nameplates, the shape of the sign system and the greening design of the dam crest roads.

3 The downstream slope decoration of the dam shall be designed in color and ornaments under the premise of ensuring the safety and stability of the dam.

4 Slope greening of large structures such as dam abutment and intake should make space for the slope planting, and measures such as vegetation concrete and thick base material spraying may be adopted. The greening of the downstream slope of the dam should be covered with lattice beams and planting troughs of horizontal berms. The thickness of the covering soil in the beams should not be less than

40 cm, and the thickness of the covering soil in the planting trough should not be less than 60 cm. The plants should be local shallow root plants that are easy to survive and maintain.

5   Under the premise of ensuring the safety and stability of the dam, the greening of the downstream weight platform shall carry out appropriate micro-topography shaping, plant shaping design, plants arrangement, etc.

6   A viewing field with parking lots shall be set up near the dam abutments if conditions permit.

6.3.2   The landscape design of the reservoir shall meet the following requirements:

1   Reservoir waterscape planning and island planning should be carried out taking into account the urban landscape needs and tourism planning along the banks of the reservoir.

2   The area upstream of the dam and the key reservoir bank areas should be designed by means of shaping varied revetment forms, creating viewing points, and adding waterfront platforms.

3   Natural and varied tree species in the form of forest belts should be adopted around the reservoir area.

4   The landscape design of the drawdown area should aim at the urban reservoir bank section, with planting measures as the main component, supplemented by appropriate engineering measures.

6.3.3   The landscape design of the operation management camp shall meet the following requirements:

1   A natural, harmonious, beautiful and comfortable living and office environment shall be created according to the environmental characteristics and functional requirements of the site.

2   The key points of landscape design shall include the shape and color of surface buildings, landscape greening in camp areas, etc., and conform to the overall style of landscape planning.

3   Necessary outdoor sports space and leisure places shall be created. The square is arranged according to the location and function of the functional area, and the design shall fully consider the construction scale of the hydropower project and the number of staff. It also needs to embody the local characteristics.

4　　The parking lots should be ecological, which can meet the basic parking function and beautify the environment.

6.3.4 The landscape design of the fish restocking station shall meet the following requirements:

1. The site selection and layout of the fish restocking station shall consider landscaping, and the landscape design shall be carried out on the basis of the determination of the fish restocking station plan.

2. The landscape layout shall be reasonably arranged according to local conditions without affecting the function, safety and structure of the station.

3. The structure forms such as outdoor pools and workshops shall meet the landscape requirements.

4. The non-construction area and sporadic space in the station shall be fully utilized for greening and beautification.

5. For qualified production workshops, building walls and slope protection sections, vertical greening requirements shall be put forward.

6. Greening planting troughs, scientific land preparation, drought resistance and water conservation measures, and auxiliary measures such as new technologies, new processes, and new materials should be adopted, to improve the survival rate of trees and the effect of landscape ecological restoration.

7. The fish restocking station shall be provided with publicity boards, signboards, picture galleries, etc. for popular science publicity.

6.3.5 The road landscape design shall meet the following requirements:

1. The road shall meet the functional requirements and create a streamlined space for driving, leisure and sightseeing.

2. The road landscape shall create a rich plant space.

3. The greening of road slopes shall follow the principles of safety and aesthetics.

4. Traffic tunnel portal landscaping shall include decorating the tunnel portal facade and greening the surrounding site.

5. The landscape design of the bridgehead of the permanent bridge should include plant landscaping, scenic spots, and publicity and sign system design.

6   Relatively flat sites, formed by road excavation or other activities, may be used as platforms for vehicle meeting, roadside parking platforms, etc., which should adopt ecological design.

6.4 Design of Landscape Architectures

6.4.1 The design of landscape architectures shall select a suitable site according to the topography and geomorphology to be in harmony with the natural environment such as landscape and vegetation.

6.4.2 The shape, style, material and color of landscape architectures shall embody regional culture and ethnic characteristics.

6.4.3 The design of landscape architectures shall meet the following requirements:

1   The form, color, theme, etc. shall be adapted to and coordinated with the surroundings.

2   The dimensions shall be determined on the ergonomic basis.

3   Chairs and stools for rest shall be provided.

6.5 Design of Environmental Facilities

6.5.1 Landscape facility ornaments may be divided into five categories according to their functions: resting ornaments, decorative ornaments, lighting ornaments, exhibition ornaments and service ornaments. Landscape facilities ornaments design shall meet the following requirements:

1   The shape, color and material shall meet human aesthetic habits and psychological needs, and embody the artistic value and cultural atmosphere.

2   The quantity, location and dimension of landscape facility ornaments shall be determined according to their main functions and human behavioral habits.

6.5.2 Resting ornaments shall include shaped garden chairs with backrests, stools, tables, etc. Natural block stones or concrete should be used to make stools and tables that imitate stone and tree stumps. When conditions permit, the low walls on the edge of flower beds may be used as chairs and stools.

6.5.3 Decorative ornaments shall include various fixed and movable flower pots, sculptures, landscape walls, landscape windows, etc., and shall be placed on selected important landscape nodes such as operation management camp, dam viewing platforms, entrance image areas, and permanent road areas of the hydropower project and designed rationally.

6.5.4 The lighting ornaments shall include the base, lamp post, lamp cap, lamps, shape, style, material and color of the garden lamp, and shall embody the regional and ethnic characteristics.

6.5.5 The display ornaments shall include publicity boards, road signs, information boards, signboards, and picture galleries. The signs shall comply with the current relevant standards of China.

6.5.6 The service ornaments shall include railings, edge decoration of flower beds and green space, and garbage bins.

6.5.7 The pavement design shall meet the following requirements:

1 The pavement design should be used for pedestrian paths, parking lots, squares, etc. in the entrance image area, operation management camp, fish restocking station, plant area, etc.

2 The pavement design shall be carried out according to the site zoning and functional requirements for activities, viewing, and rest, to divide the space, guide the sight line, and embody the artistic.

3 Pavement design shall consider color, texture and pattern.

  1) Pavement color shall be generally uniform with hue variation and be in harmony with the surroundings.

  2) Pavement texture design shall fully consider the size of the space in the selection of material texture. Paving materials with coarse texture should be used for large areas, while fine texture materials should be used for paving in subtle and key areas.

  3) The design of pavement patterns shall be novel and unique, and different patterns shall be used to beautify the environment.

4 The paving materials shall be breathable, permeable, antiskid and eco-friendly, and natural stone, wooden paving and grass-planting bricks may be selected.

6.6 Planting Design

6.6.1 The planting design shall be conducted on the overall basis from the perspectives of plan layout, function, space, scale and morphology, to meet the requirements of the overall site design for plant layout.

6.6.2 The planting design shall, based on the functional zoning of the landscape, conduct differentiation design considering the functionality and site requirements. The design shall take the requirements of the type and distribution of plant groups as the basis and embody the whole and part, unity and change, main scene and objective view, and the fundamental tree species and seasonal appearance changes, to meet the short- and long-term needs and greening effect and reduce the maintenance cost in the later period.

6.6.3 The greening plants on the dam crest and downstream weight platform of the dam complex should be shallow-rooted trees, shrubs and vines, and shall not be deep-rooted plants which might affect or damage the overall structure and functional safety of the dam. The dam abutments and the intakes and outlets should adopt 3D vegetation nets, ecological bags, geocells, vegetative blanket, and vegetation concrete for slope protection to improve the scenery effect.

6.6.4 The hydrophilous and ornamental trees and shrubs shall be reasonably mixed around the reservoir. For the drawdown area of the reservoir-bank section near key cities and towns, if the techno-economic conditions permit, the design of planting measures may be appropriately carried out.

6.6.5 The operation management camp shall select tree species and flowers with strong adaptability and ornamental value and make full use of various branches, flowers, leaves and fruits, to form a plant landscape with stable community structure and abundant seasonal appearance changes and create a good working and living environment.

6.6.6 The image exhibition area shall select local representative trees and shrubs, and adopt isolated or mass planting to display the image of the project.

6.6.7 Road greening shall comply with the current sector standard CJJ 75, *Code for Planting Planning and Design on Urban Road*, select trees with neat shape, straight trunks, lush branches and leaves, large crowns and shade, pruning resistance and high branch points as street trees, and configure shrubs and ground cover plants, to create a stereo plant landscape with rich layers and create a streamlined space for driving, leisure and sightseeing.

6.6.8 The plants around the open-air pools of the fish restocking station should be evergreen species to prevent tree leaves from polluting the water quality.

6.6.9 The planting design shall fully explore and utilize existing plant resources, and the selection of plant species shall focus on local tree species. In-situ conservation or ex-situ conservation should be adopted for the original well-grown and valuable regional representative plants, rare and protected plants, old and notable trees in the project area and reservoir-inundation area. The planting design shall give priority to using the topsoil stripped from the construction area and the reservoir-inundation area as the planting soil.

6.7 Landscape Lighting Design

6.7.1 The landscape lighting design of a hydropower project shall focus on functional lighting, supplemented by ornamental lighting. Landscape and decorative lighting design shall consider the impact on pedestrians, plants and the surroundings. Landscape lighting design shall comply with the current sector standard JGJ/T 163, *Code for Lighting Design of Urban Nightscape*.

6.7.2 The landscape lighting design shall consider ecological and environmental protection requirements, and reasonably determine the illumination angle, mounting location and shading measures to avoid the impact of light pollution.

6.7.3 The shape, mounting height and location of various lamps shall be in harmony with the overall landscape, and the integration of regional culture or corporate culture should be considered.

6.7.4 High-efficiency and energy-saving lamps shall be selected, and solar lamps should be used in areas where conditions permit.

6.7.5 The lighting lines shall be coordinated with other pipelines in the area to avoid interference and repeated construction.

6.8 Design of Auxiliary Facilities

6.8.1 Auxiliary facilities shall include management buildings, water supply facilities, power supply facilities, parking lots, docks, public toilets, etc. The design of auxiliary facilities shall meet the following requirements:

1. The design of auxiliary facilities shall be forward-looking and meet the needs of short-term planning, construction and long-term development.

2. On the premise of satisfying the basic functions, the auxiliary facilities should continue the style of main buildings and structures to increase the landscape heterogeneity, diversity and stability.

3. The design area, capacity, service radius, and indicators of auxiliary facilities shall reserve space and create conditions for later tourism planning and development.

6.8.2 The design of security facilities shall meet the following requirements:

1. The design of security facilities shall be guided by the security facility planning and comply with the current relevant technical standards.

2. The architectural design of security facilities such as guard rooms, fences, and railings shall be in harmony with the overall appearance of the hydropower project.

3   The monitoring and control system pipelines shall be coordinated with other pipelines in the area to avoid interference and repeated construction.

6.9 Design Outcomes

6.9.1 The landscape design for a hydropower project shall propose the design outcomes according to the requirements of the scheme design, preliminary design and construction drawing design, respectively.

6.9.2 The scheme design documents shall include design report, design drawings and cost estimates, and shall meet the following requirements:

1   The design report shall include design basis and basic information, project overview, general layout design, zone layout design, etc.

2   The design drawings shall include the general plan, functional zone plan, general plan of planting design, enlarged plan of main scenic spots, elevations or renderings of main scenic spots, etc.

6.9.3 The preliminary design documents shall include design report, design drawings and cost estimates, and shall meet the following requirements:

1   The design report shall include design basis and basic information, project overview, general layout design, elevation design, planting design, main waterscape design, main landscape architecture design, main landscape ornament design, pavement design, techno-economic indicators, etc.

2   Design drawings shall include general drawings, zone drawings, local important node drawings, landscape architecture drawings, planting drawings, waterscape drawings, associated landscape drawings, etc.

3   The general drawings shall include the general setting out, general index drawing, general elevation and general planting drawing.

4   The zone drawings shall include zone layout drawings, zone index drawings, zone elevations, and zone planting drawings.

5   The local important node drawings shall include detailed drawings of pavement.

6   The landscape architecture drawings shall include plan, elevation and section drawings.

7   The waterscape drawings shall include still water drawings, moving water drawings, etc.

8   The associated landscape drawings shall include plant irrigation

drawings, landscape water supply and drainage drawings, landscape lighting design drawings, etc.

6.9.4 The construction drawing design documents shall include design report and design drawings, and shall meet the following requirements:

1. The design report shall include design basis, project overview, material description, waterproofing and moistureproofing description, planting design description, etc.

2. The design drawings shall include general drawings, zone drawings, local important node drawings, landscape architecture drawings, planting drawings, waterscape drawings, associated landscape drawings, etc.

3. The general drawings shall include the general line drawing, general index drawing, general elevation and general planting drawing.

4. The zone drawings shall include the zone drawings, zone index drawings, zone elevations, and zone planting drawings.

5. The local important node drawings shall include detailed drawings of pavement and construction methods.

6. The landscape architecture drawings shall include plan, elevation and section drawings, structural drawings, and detailed drawings of construction methods.

7. The planting drawings shall include zone plant arrangement and seedlings list.

8. The waterscape drawings shall include still water drawings, moving water drawings, etc.

9. The associated landscape drawings shall include plant irrigation drawings, landscape water supply and drainage drawings, landscape lighting design drawings, etc.

7 Costs

7.0.1 The landscape investment calculation for a hydropower project shall follow the principle of "no repeating and no missing", and the costs for main works and ecological protection works shall not be repeatedly counted.

7.0.2 The cost preparation for the landscape of a hydropower project shall be in accordance with relevant current quotas and specifications for hydropower projects and soil and water conservation works. If the preparation requirements are not specified, the cost preparation may be in accordance with relevant sector quotas and specifications issued by the state or local authority.

7.0.3 The cost of landscape works of a hydropower project in planning and design shall include construction and installation works cost, separate cost and contingency. The composition of each cost shall meet the following requirements:

1. Construction and installation works cost shall include direct cost, indirect cost, profit and tax.

2. Separate cost shall include construction management cost, operational production cost, investigation and design cost, other taxes and fees.

3. The contingency shall include the basic contingency and contingency for price variation.

Appendix A Indicators and Evaluation Standards for Landscape Planning of Hydropower Projects

A.0.1 The landscape planning indicators and evaluation standards for hydropower projects shall comply with Table A.0.1. The landscape planning effect evaluation, for which the full mark is 120, shall be independently scored by at least 10 people and the average shall be taken.

Table A.0.1 Landscape planning indicators and evaluation standards for hydropower projects

Level 1 indicators /full marks	Level 2 indicators /full marks	Level 3 indicators /full marks	Evaluation standards	Score
Landscape resource value/50	Ecological landscape value/20	Waterscape/5	The water has a high landscape beauty: the water surface is 10 km^2 or above, and the water body types are very rich and the water quality is very good	5
			The water has a medium landscape beauty: the water surface is 5 km^2 to 10 km^2, and the water body types are rich and the water quality is good	3
			The water has a lower landscape beauty: the water surface is 5 km^2 or below, and the water body type is single and the water quality is mediocre	1
		Native geomorphologic landscape/5	The terrain is very unique and the landform types are very rich	5
			The terrain is unique and the landform types are rich	3
			The terrain is common	1
		Biodiversity/5	The vegetation coverage is very high, flora and fauna are very rich, and there are rare species	5

Table A.0.1 *(continued)*

Level 1 indicators /full marks	Level 2 indicators /full marks	Level 3 indicators /full marks	Evaluation standards	Score
Landscape resource value/50	Ecological landscape value/20	Biodiversity/5	The vegetation coverage is high, and the flora and fauna are rich	3
			The vegetation coverage and flora and fauna are moderate	1
		Landscape pattern/5	The landscape continuity is very good, the integrity is very high, and the ornamental value is very good	5
			The landscape continuity is good, the integrity is high, and the ornamental value is good	3
			The landscape continuity, integrity, and ornamental value are common	1
	Cultural landscape value/10	Surrounding scenic resources/5	Surrounded by national scenic spots	5
			Surrounded by provincial scenic spots	3
			Surrounded by municipal and county-level scenic spots	1
		Historic sites and cultural customs/5	There are historical relics, old villages, precious monuments or intangible cultural heritage around the project area	5
			There are historical and old villages around the project area, with rich folk culture and customs	3
			There are historical villages around the project area	1

Table A.0.1 *(continued)*

Level 1 indicators /full marks	Level 2 indicators /full marks	Level 3 indicators /full marks	Evaluation standards	Score
Landscape resource value/50	Project landscape /20	Engineering representation /5	The uniqueness is very obvious, the popularity is well known, and the industry influence is great	5
			The popularity is known, and the industry influence is large	3
			The popularity and industry influence are moderate	1
		Architectural art effect /5	The dam and powerhouse are highly ornamental	5
			The dam and powerhouse are ornamental	3
			The dam and powerhouse are somewhat ornamental	1
		Science and technology /5	Be environment-friendly, there are numerous technological innovations, and the level of intelligent management is very high	5
			Be environment-friendly, there are many technological innovations, and the level of intelligent management is high	3
			Be environment-friendly, there are few technological innovations, and the level of intelligent management is moderate	1
		Garden ornamental value /5	The landscape continuity is very good, the integrity is very high, and the ornamental value is very high	5

Table A.0.1 *(continued)*

Level 1 indicators /full marks	Level 2 indicators /full marks	Level 3 indicators /full marks	Evaluation standards	Score
Landscape resource value/50	Project landscape /20	Garden ornamental value /5	The landscape continuity is good, the integrity is high, and the ornamental value is high	3
			The landscape continuity, integrity and ornamental value are common	1
Development and utilization conditions /30	Traffic conditions /10	Location /5	Within 50 km from large and medium cities, counties or world and national scenic spots	5
			50 km to 200 km away from large and medium cities, counties or world and national scenic spots	3
			More than 200 km away from large and medium cities, counties or world and national scenic spots	1
		Accessibility /5	Highway	5
			First level road	3
			Road below the second level	1
	Basic service facility conditions /15	Infrastructure /5	The water, electricity, communication and other infrastructure equipment is well-equipped and operating normally	5
			The water, electricity, communication and other infrastructure equipment is provided and operating normally	3
			The water, electricity, communication and other infrastructure equipment is inadequately provided and operating normally	1

Table A.0.1 *(continued)*

Level 1 indicators /full marks	Level 2 indicators /full marks	Level 3 indicators /full marks	Evaluation standards	Score
Development and utilization conditions /30	Basic service facility conditions /15	Reception facilities /5	Within 20 km of the project, there are three-star hotels or hotels with hardware conditions of above three-star	5
			Within 20 km of the project, there are two-star hotels or hotels with hardware conditions of above two-star	3
			There are other facilities	1
		Recreational facilities /5	There are many types of good recreational facilities	5
			There are few types of recreational facilities	3
			There are very few types of recreational facilities, and the supporting facilities are not perfect	1
	Environmental carrying capacity /5	Annual capacity /5	100000 and above	5
			50000 to 100000	3
			50000 and below	1
Planning effects /40	Visual impact /25	Scenery intensity /5	The landscape design within the scope of the project has rich vegetation types, harmonious colors, excellent materials, and high ornamental value	5
			The landscape design within the scope of the project has few vegetation types, harmonious colors, ordinary materials, and intermediate ornamental value	3

Table A.0.1 *(continued)*

Level 1 indicators /full marks	Level 2 indicators /full marks	Level 3 indicators /full marks	Evaluation standards	Score
Planning effects /40	Visual impact /25	Scenery intensity /5	The landscape design within the scope of the project has few vegetation types, disharmonious colors, poor materials, and low ornamental value	1
		Landscape recognition /5	The landscape has a high recognition and is distinctive, with unique in the region	5
			The landscape recognition is satisfactory, is similar to other landscapes, and has certain characteristics of its own	3
			Landscape recognition is ordinary, relatively common, and very similar to other landscapes	1
		Cultural integration /5	The landscape plan within the project scope has cultural connotations in general and is highly integrated with regional culture	5
			The landscape plan within the project scope reflects cultural elements in local design and adopts cultural symbols in detail design	3
			The landscape plan within the project scope fails to reflect the local cultural content	1
		Landscape spatial richness /5	The landscape spatial design within the project scope for private space, semi-private space and public space has a reasonable hierarchy, with rich types of distribution, rest and viewing	5

Table A.0.1 *(continued)*

Level 1 indicators /full marks	Level 2 indicators /full marks	Level 3 indicators /full marks	Evaluation standards	Score
Planning effects /40	Visual impact /25	Landscape spatial richness /5	The landscape space design within the project scope has a certain combination and hierarchy	3
			The landscape spatial design within the project scope is relatively simple	1
		Compatibility with the main works /5	The planned landscape effect integrates and echoes with the main works of the hydropower project and the water conservancy structures, forming a harmonious environment	5
			The planning landscape effect integrate to some extent with the main works of the hydropower project and the water conservancy structures	3
			The planned landscape effect integrates little with the main works of the hydropower project and the water conservancy structures	1
	Functional perception /15	Layout reasonableness /5	The planned landscape has reasonable zoning and spatial structures, beautiful layout, and rich tour routes that are conveniently accessible to each other	5
			The planned landscape has reasonable zoning and spatial structures, clear layout, and convenient tour routes that are accessible to each other	3

Table A.0.1 *(continued)*

Level 1 indicators /full marks	Level 2 indicators /full marks	Level 3 indicators /full marks	Evaluation standards	Score
Planning effects /40	Functional perception /15	Layout reasonableness /5	The planned landscape has reasonable zoning and spatial structures, ordinary layout, and reasonable tour routes	1
		Landscape service facilities /5	The planned landscape has complete and diverse supporting facilities, fully considering the needs of environmental psychology, ergonomics and aesthetics	5
			The planned landscape has complete supporting facilities, meeting the needs of public services, fully considering the needs of ergonomics, and possessing a certain degree of psychological and aesthetic functions	3
			The planned landscape has complete supporting facilities, basically meeting the needs of public services	1
		Landscape structure configuration /5	The landscape structures have appropriate dimensions, are harmonious with the overall environment, and have a certain unique style and cultural decorations	5
			The landscape structures have reasonable dimensions and their appearances and colors conform to the overall style and are not obtrusive	3

Table A.0.1 *(continued)*

Level 1 indicators /full marks	Level 2 indicators /full marks	Level 3 indicators /full marks	Evaluation standards	Score
Planning effects /40	Functional perception /15	Landscape structure configuration /5	The dimensions and style of the landscape structures are not closely integrated with the overall environment	1
total score				

NOTE The "water body type" refers to the combination of multiple water body forms, which is considered "very rich" if there are three types or more, "rich" if there are two types, and "single" if there is one type.

A.0.2 The landscape planning effect evaluation for hydropower projects shall be divided into three grades, and the grading criteria shall comply with Table A.0.2.

Table A.0.2 Grading criteria for evaluating the effect of landscape planning for hydropower projects

No.	Grade	Grading criterion
1	I	$S \geq 95$
2	II	$85 \leq S < 95$
3	III	$70 \leq S < 85$

NOTE S is the total score of the evaluation.

Appendix B Contents of Landscape Planning Report for Hydropower Projects

1 Planning Background
1.1 Geographical Location
1.2 Natural Environment
1.3 Social Economy
1.4 History and Culture
1.5 Related Plans

2 Analysis of the Current Situation
2.1 Project Overview
2.2 Site Analysis

3 Planning Basis and Principles
3.1 Planning Basis
3.2 Planning Principles

4 Planning Objectives and Concepts
4.1 Planning Objectives
4.2 Planning Indicators
4.3 Planning Concepts

5 Planning Scheme
5.1 General Layout
5.2 Functional Zoning
5.3 Zone Planning

6 Special Planning
6.1 Plant Arrangement Planning
6.2 Landscape Lighting Planning
6.3 Traffic Streamlines Planning
6.4 Tourism Development Planning
6.5 Construction Sequence Planning

7 Cost Estimation

Explanation of Wording in This Code

1   Words used for different degrees of strictness are explained as follows in order to mark the differences in executing the requirements in this code.

   1) Words denoting a very strict or mandatory requirement:

   "Must" is used for affirmation; "must not" for negation.

   2) Words denoting a strict requirement under normal conditions:

   "Shall" is used for affirmation; "shall not" for negation.

   3) Words denoting a permission of a slight choice or an indication of the most suitable choice when conditions permit:

   "Should" is used for affirmation; "should not" for negation.

   4) "May" is used to express the option available, sometimes with the conditional permit.

2   "Shall meet the requirements of…" or "shall comply with…" is used in this code to indicate that it is necessary to comply with the requirements stipulated in other relative standards and codes.

List of Quoted Standards

CJJ 75, *Code for Planting Planning and Design on Urban Road*

JGJ/T 163, *Code for Lighting Design of Urban Nightscape*